Arms for Oil

Michael Barratt Brown

SPOKESMAN
For
SOCIALIST RENEWAL

First published in 2005 by
Spokesman
Russell House, Bulwell Lane
Nottingham
NG6 0BT
Phone 0115 970 8318   Fax 0115 942 0433.
e-mail elfeuro@compuserve.com
www.spokesmanbooks.com
©Michael Barratt Brown

ISBN 0 85124 698 2

A CIP catalogue is available from the British Library.

Printed by the Russell Press Ltd (phone 0115 978 4505)

Arms for Oil

Michael Barratt Brown

'Part II of the Al Yamamah deal for Tornado strike aircraft, BAe Hawk trainer
jets, Sandown class mine hunters and Black Hawk helicopters together worth
some £40 bn… called for a £1.5 bn. down payment in cash, the rest in oil, a
deal designed by Peter Levene of the MoD [UK Ministry of Defence], calling
for payment of as much as 500,000 barrels of oil per day, which were then to
be traded by Shell and BP.' (Sunday Times, 9.10.94, quoted in Gerald
James, In the Public Interest, Little, Brown, 1995)

It is a fact, an unfortunate fact, that economic relations between the
West and the Arab and other Islamic countries could for many years
be described accurately in just three words 'Oil for Arms' or
alternatively 'Arms for Oil'. This has a long history going back to the
First World War, when the Ottoman Empire was broken up between
the British and French Empires, and client regimes were established
in the newly created states. The Second World War brought in the
American Empire, but the main purpose of all these empires was to
control the rich oil reserves in the region. Arms were supplied as
necessary to maintain the client regimes in power. This became
essential if oil was to remain cheap.

Energy policy has for long been a highly debatable issue in the
United States and in Western Europe. These regions in the West had
large reserves of coal, and the US had oil (Table 2). Until the North Sea
oil and gas were discovered and oil prices rose high enough for oil to
be worth exploiting there, Western Europe had only coal, which could

NOTE: I am having to use the words 'Middle East' to describe certain mainly
Arab Islamic countries, because this is how they appear in UN statistical
publications, often together with North Africa. The countries so described are
listed in Appendix Table 1, attached to this paper. As they include Israel this
does distort the picture and where possible I have tried to distinguish Israel
from the figures which I present. Some of the UN statistics do separate the
Arab States and I use these when appropriate. I do, however, write the words
'Middle East' at all times in quotation marks in order to emphasise that such
a Euro-centred definition is not my own.

be converted into petroleum but at a cost. For the needs of cars and aeroplanes and increasingly for ships and railway engines, oil had to be imported. The issue was always about price. Even in the United States the costs of extracting oil rose as the largest and most accessible indigenous oil fields were worked out. The availability of cheap oil from the 'Middle East' became a most attractive option for the USA as much as for the other industrialised countries which had no oil (Table 3).

There was from the start strong Western government involvement in the control of 'Middle East' oil supplies. The UK's original interest early in the Twentieth Century was in fuel for the British navy, and BP was established as a state owned company, and is still under virtual government control. In February 1945, President Roosevelt signed an agreement on board the USS *Quincy* in the Suez Canal with King Abdul Aziz of Saudi Arabia which established the US-Saudi relationship based on the exchange of oil for security. The United States government held a majority shareholding in ARAMCO, which came to behave like a state within a state in Saudi Arabia. ARAMCO built a pipeline to carry Saudi oil across Jordan and Syria to the Lebanese port of Sidon, and proceeded to create large reserves of oil in a Petroleum Reserve Corporation, which enabled the United States for long to control the world price of oil. The price paid to the producers was held at about one sixteenth of the selling price of crude oil. According to David Harvey in *The New Imperialism* (Oxford University Press 2003, p.20), between 1940 and 1967 US companies increased their control of 'Middle East' oil reserves from 10% to 60%, reducing British control from 72% to 30%.

In the 1950s, oil became increasingly the chief source of energy in the West. Even those countries like the United Kingdom, which had large reserves of coal, came to choose oil and gas, not only for transport but for heating and lighting. This was because these fuels were easier to use, and above all because they were cheaper. Costs of minerals depend on many factors: the machinery and labour required for extracting them from the earth, for transport and for refining and processing, and there may be hidden costs. For, there is also the question of security of supply. Many countries' governments protect their own supplies with subsidies to keep the suppliers happy, and to ensure continuity and security of supply. Governments for this reason

2

protect their food producers and also their coal miners, but there are limits to public support for such open subsidies. They are rather more obvious than the hidden subsidies of the military costs of controlling foreign supplies. Such possible military costs were hardly considered when the coal industries of Western Europe were closed down in the 1970s. Once any country adopted imported oil and gas as the main sources of supply of energy, such supplies had then to be kept cheap, and that meant exercising control over the resources and the security of supply.

In this paper it will be necessary to examine the hidden costs of reliance on cheap oil and gas by those countries which took that decision. The several wars fought over control of the supplies of oil in the 'Middle East' and North Africa bear witness to the importance of these sources of supply. For the simple fact is that this region contains by far the largest and richest reserves in the world of potentially cheap oil and gas. To these must now be added the reserves of oil and gas surrounding the Caspian Sea, but throughout most of the 20th Century fossil fuel reserves in the 'Middle East' and North Africa amounted to about two thirds of all easily recoverable reserves of oil and a quarter of all natural gas (Table 2). It is an interesting gift of Nature that the new fossil fuel reserves are once again found, as also in Nigeria and Indonesia, under lands inhabited by peoples holding the Islamic faith.

Not only does the 'Middle East' region contain the richest reserves of cheap oil and gas, but the oil and gas in the other regions of current supply are due to run out during the second decade of the 21st century. This is true of North America, the North Sea, Africa, Russia and China. In the absence of alternative energy sources, competition for access to the remaining reserves of cheap oil and gas is likely to become stiff. Hence the alarming title of a book written as early as 1979 by P.Peteers, *Can We Avoid A Third World War Around 2010?* The war he anticipated would be over control of the remaining fossil fuels. Caspian oil and gas reserves were not recognised in 1979, so that up till that time the 'Middle East' alone had reserves with a life of some fifty years into the new century. It was United States anxiety on this account that probably led to the establishment by the US in the 1980s of a 'Gulf Cooperation Council' with Saudi Arabia, Kuwait, the United

Arab Emirates and other Gulf states. Military equipment was offered as a back-up to US military forces to be stationed in the region. Even before the first Gulf War, several thousand US military personnel were positioned in several of the Gulf states, and the Gulf War created the need for a massive US base in Saudi Arabia itself, which came to be deeply resented by such Islamic fundamentalists as Osama Bin Laden

Oil Supplies and Oil Prices

'Middle East' and North African reserves of fossil fuels should have made the oil states in the region enormously rich and the centres of advanced industrialisation. This was not, however, the purpose of the imperial powers. Almost all the oil was exported as crude petroleum for refining and converting into oil based products in the imperial centres themselves. Even today less than 10% of oil and natural gas exports from the region consist of refined petroleum and other oil products. Sales of crude petroleum make up around 80% to 90% of the total exports of the 'Middle East' and North African countries (Table 5). Although their combined populations amount to less then 400 million out of a world total of 6000 million (7%), their oil exports make up about 50% of all world oil exports and two thirds of those from outside the United States, Europe and China (Table 4).

The dollar value of annual oil exports, at more than $500 billions, is equal to the combined value of all other primary commodities entering world trade. Even here it is necessary to recognise that the figures in dollar values for oil sales in any year can be misleading. This is for two reasons. First, some oil is supplied under barter arrangements such as the 1985 Al Yamamah three part deal between the United Kingdom and Saudi Arabia, a deal worth altogether some £120 billions, quoted in the epigraph at the head of this paper. Another example was the purchase by Prince Sultan of Saudi Arabia of 10 Boeing 747s for Saudi Airlines in exchange for 34.5 million barrels of oil. Second, it has to be emphasised that the purpose of the imperial powers was not only to control the oil supplies and use them for developing their own industries and transport systems, but also to get the oil cheap. This was achieved both by the imperial relations established with the rulers of oil producing countries and by a cartel of the world's main oil companies, the so-called 'Seven Sisters' as they were before amalgamations – Exxon, Mobil, Chevron, Gulf, Texaco, Shell and British Petroleum

These giant oil companies are among the largest companies in the top twenty of the world's transnational corporations. From the start of oil exploration they took over the drilling, transporting, refining, processing and, finally, retailing the crude material. They refrained from competing with each other in their purchases and relied on imperial arms for maintaining in the oil producing countries such regimes as would not push for higher prices. Instead, these regimes could live in grand style on the royalties they were able to pocket individually from oil sales. The oil companies' aim was to maximise their own profits, but this involved arrogating to themselves sovereign rights – where to explore for oil, how much of their profits to invest in exploration, how much oil to produce, what price to charge, how to share the proceeds among themselves, how to transport the oil, and what regimes in the oil producing countries to support or dispense with. All these measures in effect deprived the oil producing states of the greater part of the value of their oil.

Resentment among the oil states at the foreign exploitation of their natural assets grew during the 1950s. In 1951, the newly elected Prime Minister of Iran, Muhammad Mossadeq, nationalised the Anglo-Iranian Oil company, and three years later newly elected Egyptian President Gamal Abdel Nasser nationalised the Suez Canal and declared Saudi oil to be Arab oil. The British and French tried but failed to uproot Nasser at Suez in 1956, having omitted to consult the United States, but a combined US CIA and UK MI6 operation overthrew Mossadeq. The US government through ARAMCO had negotiated an agreement with Saudi Arabia providing for the Saudis to obtain 50% of the price of a barrel of oil after deducting production, transport and marketing costs; not much of a deal for the Saudis, but better than the other oil states were getting. World crude oil prices remained fixed from 1948 to 1958. ARAMCO's profits still rose in these years by 300%.

That was the position until 1959 when, under the leadership of Saudi Oil Minister Abdallah Tariki, and with the support of the Venezuelan Oil Minister Perez Alfonso, discussions began about forming an oil producers' cartel. It took another three years to persuade the other oil states to form the Oil Producers' Executive Council (OPEC). The founding meeting in 1962 was hosted by Iraq, and designed to outface the 'Seven Sisters'. The result was not long delayed. The 'unfriendly' Iraqi government of Abdel Karim Kassem was overthrown, and one friendlier to the West, led by Saddam Hussein, put in its place. Similar action taken by combined US and UK intelligence in Iran had restored the Shah to the throne, but the Shah fell in 1979 to an Islamic revolt. This led to a switch of US/UK support from Iran to Iraq, and the encouragement of Saddam Hussein to invade Iran. The United Nations placed an embargo on arms sales to both sides, which the United States, United Kingdom, France and Russia all sought to breach by various subterfuges, which will be the subject of later discussion in this paper.

For some years after 1962 OPEC had only a small effect on oil prices. These slowly increased year by year until they had doubled by the time of the Israeli-Egyptian war in 1967. When, however, that war was resumed in 1973 with US armed support for Israel, the OPEC members at length determined to play the oil card and imposed an oil embargo. Oil prices soared – to twenty times the 1967 price by 1974, a hundred times the 20 cents a barrel at which they had stayed for so long despite inflation of other prices, since the end of the Second World War. Even though, under pressure from the imperial powers, the embargo was lifted within a year, oil prices continued to be kept high by restrictions on supply, and leapt up again with the Iran-Iraq war in 1980 (Table 6), and again in the Iraq war of 2003-4.

An important economic result of the oil price hike in the 1970s and 1980s was that huge surpluses of oil money flowed into the world's financial system. Oil had since the 1940s been traded for dollars, so that the payments' balances on the foreign trade of the USA could always be augmented by the oil money. One of the actions of Saddam Hussein in the early 21st century most offensive to the United States was that he began to trade Iraq's oil for Euros. The oil money which flooded into world markets in the 1970s not only benefited the United States, but came to be used by US and UK banks and other banks in London as Euro-dollars to finance a great expansion of world trade. After some years the disadvantages of high oil prices for US businesses and petrol consumers led to heavy pressure on Saudi Arabia to bring the price down. Prices rose again with the outbreak, in 1980, of the Iran-Iraq war, despite the efforts of the Saudis to keep prices down by pumping out more oil. The price of crude petroleum from 1981-5 averaged about $30 a barrel, but in 1986 it fell to less than half that figure, still well above the price in the 1950s and 60s. Oil prices rose again temporarily in the first Gulf War and once more in the second (Table 6). Most recently, as a result of unsettled conditions in Iraq, the price has reached more than $50 a barrel.

The power of OPEC has been much reduced through the development of Alaskan and North Sea oil and gas and the emergence of many smaller oil producers who have not joined OPEC. These developments were all made viable by the higher oil prices after 1973. There are considerable resources of oil and tars available in the earth's crust, but many are costly to extract and extraction therefore depends on the price at which they can be sold. The great importance of the sources of oil in the 'Middle East' is that the oil is easy to extract. It is therefore in the interest of the producers in this region, and particularly of the Saudis who have the largest oil fields, to keep world prices low. As this is also beneficial to the chief oil consuming countries, a situation of interdependence has developed. Oil prices only rise in times of serious threat to supplies, which means in effect in times of war, as has just been indicated. Such large variations followed from fears of the reduced availability of oil when wars were raging in the 'Middle East', but the continuing lower level of prices between wars has been maintained not only by Saudi intervention but by the competition of increased use of natural gas and by some rather limited measures of conservation. Much of the gas comes from the same sources as the oil, but Russia has exceptionally large gas reserves.

The variations in price inevitably have a marked effect on incomes in the oil producing countries and on their foreign balance of payments. This last could vary in the case of Saudi Arabia from a surplus of $50 billions a year in the late 1970s and early 1980s to a deficit through the rest of the 1980s and early 1990s. (Table 6). Partly because of difficulties in restarting oil supplies from Iraq after the first Iraq War, and even more since the second Iraq war, and partly because of Saudi discontent with US 'Middle East' policies in general, oil prices have stayed up. Saudi financial surpluses have thus been restored. Such surpluses make possible investment in financial and other foreign assets, or an increase in other foreign purchases – of arms for example.

Where the Oil Wealth Goes

It is clear from the exercise of imperial power over the oil states that only a very small part of the value added to the crude material remains with these states. The greater part goes to the giant Western oil companies, who sell the petrol, plastics, fertilisers and other petroleum products. From their sales governments in the West derive large revenue in the form of carbon taxes and seigneurial rights deriving from the sale of oil in their currencies. The Western populations and particularly the American people still enjoy cheap fuel oil for their cars and aeroplanes, although pollution and traffic congestion somewhat reduce this enjoyment.

Even this list is an understatement of the benefits accruing to the West from control over the refining, distribution and processing of crude petroleum. Political power in the West, and most particularly in the United States, is not unconnected with the wealth derived from oil sales. The oil companies are heavily involved in financing the campaigns of the two rival political parties in the United States, and the Bush family themselves made their money in oil, while the chairman of BP is generally to be found among the leading advisers of British governments, not excluding New Labour. Since BP was first established by a British government to supply the Royal Navy, the British government still holds a major share or has hidden control. At the same time, the Saudi oil princes are not above providing funds for US election expenses. Ronald Reagan's second election campaign is quoted as an example, and there may be others (*The Guardian* Special Report 31.03.04, page 4).

What is left for the peoples of the oil states is not small, though subject to the ups and downs of the oil prices. Annual income per head in Saudi Arabia in 2001 was estimated at $13,330, not quite back to the 1980 figure of $14,600, but well above $6,000 in 1990. The average per capita income of all the Arab states was just over $5,000 in 2000, but the annual growth rate was only 0.3% per year over the 15 years from 1975, a slightly better 0.7% from 1990 (Table 1). Compare the annual incomes of the countries with the highest development at some $25,000 with annual increases of over 2%. The obvious failure of the oil producing states lies in the absence of the development of a manufacturing industry. There was in fact some development between 1980 and 1990 – from only 9% of GDP in manufacturing for all West Asian countries in 1980 to 16% in 1990, but the share of manufacturing thereafter actually fell back – to 15% in 2001. In Saudi Arabia, Oman and the United Arab Emirates the share rose to around 10%, but this has to be compared with figures of 25% in the rest of the Developing Asian countries (Table 7).

Those Developing Countries which had no oil and failed to industrialise suffered from the oil price hikes, especially that in the early 1980s. Their oil imports cost more, while the prices of their own commodity exports fell steadily, many of them being in long term competition with the synthetic materials derived from oil products. These countries ran up debts with the encouragement of the Western banks flush with oil money. Interest rates rose with rising inflation and the debts accumulated. 'Middle East' and North African countries with little or no oil – Morocco, Tunisia, Jordan and Egypt – suffered in this way, and they suffered particularly because the international financial institutions required them to increase their commodity exports to repay their debts. As all commodity producer countries were given the same advice, stocks of their staple exports built up and prices fell still further. The result of this bad advice was that most of such non-oil producing countries in the 'Middle East' and North Africa failed to develop manufacturing industries and to switch their exports to manufactured goods. Turkey and Israel were the exceptions.

Free trade is not necessarily fair trade. The artificial world division of labour established by colonial rule, leaving the colonies to produce the raw materials for the industrialisation of the colonial powers, is self-perpetuating. Breaking out of it is very difficult for two reasons. First, it is hard for one-time colonies to make their infant industries competitive in free trade with already industrialised economies. Second, ruling élites become established in colonies with an interest in the trading of their raw materials and easily open to corruption by the one-time colonial powers. Colonies which had no oil were in the weakest position, having the least bargaining power in world trade, and generally subject to the exploitation of giant corporations from the West and from the middle-men who act as their intermediaries. Non-oil producing African countries suffered worst (see M. Barratt Brown, *Africa's Choices – After 30 Years of the World Bank*, Penguin, 1995)

What is more surprising, the oil producing countries have also failed to make the switch to manufacturing, when they had ample resources to do so. Both the reasons which have just been noticed apply to the oil states – both the obstacles facing infant industries and the resistance of existing colonial élites benefiting from the colonial trade. Even when some manufacturing is established in one-time colonies, such development is hard to maintain. Comparing the world rankings in production of manufactures per head of population, and also the rankings of manufactured exports world wide, the big oil producers – Bahrein, Saudi Arabia and Algeria – all actually fell back sharply in such rankings between 1985 and 1998 (Table 8). This is a seriously retrograde development, which was noted in the UNIDO *Industrial Development Report* for 2002-3. UNIDO's review of 'Global Industrial Activity for the Middle East and North Africa' concluded that this region

> 'has achieved a fair manufacturing value added per capita, a reasonable base of skills and infrastructure and good access to foreign technology. But its industrial and export structures are not geared to technology upgrading and its technological effort is weak.'

One has to assume that this is the result of the particular relations established in the exchange of oil for imports from the industrialised West. It does not appear that the ruling élites of the oil states can be entirely blamed for this failure to use their rich resources for diversifying development and for winning more of the value added to their oil exports. On at least one occasion – in the 1950s – King Saud formed a partnership with the Greek ship owner, Aristide Onassis, to create a tanker fleet – the Saudi Maritime Tanker Company – to transport his country's oil. The US oil companies at once turned to their government to resist this development. The US ambassador to Saudi Arabia and the US Secretary of State himself – John Foster Dulles – saw this as a first step to nationalisation, following the example of the Egyptian leader Gamal Abdel Nasser, and issued specific threats to stop all US oil imports from Saudi Arabia. After that, according to the king's biographer, Robert Lacey (*The Kingdom*), the Saud family decided to obey their imperial masters. Others such as Mossadeq in Iran and Kassem in Iraq disobeyed, and suffered the consequences.

They had little or no help from public opinion in the West, for whom cheap oil seems of paramount importance. There is a Fair Trade Movement in the United Kingdom and in most Western countries, which seeks to build a fairer exchange in trade with raw material producing countries. This is co-ordinated by an International Fair Trade Association and a Fair Trade Labelling Organisation to establish fair trading criteria. Fair trade coffee, chocolate, nuts, bananas and other tropical fruits are to be found in the big super markets. Some fair trade cotton goods and many 'ethnic' craft products are on sale in specialist shops. An attempt to import olive oil from Palestine to the United Kingdom on fair trade terms was sabotaged by the Israeli authorities. Imports of minerals have proved still harder to achieve on fair trade terms, although Cuban nickel and Nicaraguan gold were for some years imported by one fair trade organisation. Entering the oil market would require more capital than Fair Trade organisations can at present hope to accumulate.

The major factor in the use of 'Middle East' oil wealth is, however, that it is most unequally distributed inside the producing countries. This in itself holds back development, since poor producers make for poor markets. According to published sources, there are over 50 billionaire princes in Saudi Arabia, but they buy mainly from the West their planes and cars and clothes and luxuries. Although the value of each of their palaces is measured in billions of US dollars, most of their wealth is spent outside the country. The great wealth of a few must imply much inequality in the distribution of the average Saudi Arabian income per head. 3% of the population are recorded by the United Nations Development Programme as being undernourished, and 14% of children are under-weight up to the age of five. In the Arab States as a whole 13% were said to be under weight in 2001 (Table 9). Figures for income distribution and the incidence of poverty are not published for most of the 'Middle East' countries in

11

either the World Bank or UNDP reports. But, apart from under-nourishment, some other clues are available.

The urban population has grown rapidly throughout the 'Middle East' and North Africa, bringing with it access to clean drinking water and proper sewerage in the towns (Table 1). As a result, infant mortality rates in these countries were reduced between 1970 and 2000 from 128 per 1000 children under five years to 49 per 1000. This must still be compared with 6 per 1000 in the West. Life expectancy remains at around 66/69 years (male/female) compared with 10 years longer in the West (Table 9). Public expenditure on health in the Arab States was raised in the last decade of the century as a proportion of national income (GDP), from under 2% to an average of about 2.5% (over 4% in Saudi Arabia). This is still half of the proportion of GDP spent on health in the United Kingdom or the United States (Table 10). The UN Global Report on Human Settlement 2003, *The Challenge of Slums* (page 113), places North Africa and the 'Middle East' among the middle rank of Developing Countries in respect of housing, water connection, sewerage and electricity supply. This is much lower than might be expected from the average national income per head.

An improvement in the share of public expenditure going to education was also recorded throughout the Arab States in the last decade of the Twentieth Century, the proportion of GDP spent on public education having risen from around 4% to nearly 4.5%, which is near to the proportion spent on public education in the United States and United Kingdom (Table 10). Saudi Arabia and Yemen have shown particularly high figures of education spending, reaching around 10% of GDP in the year 2000. Adult literacy is claimed for about 60% of the Arab States' populations over 15 years of age, compared with 50% in 1970, but still well below almost universal literacy in the West. As a result of the increase in public spending in Saudi Arabia, this country achieved a literacy rate of 77% compared with 66% in 1990. Access to primary education is available to 77% of children in the Arab States, an advance on 73% in 1990, and not much less for secondary education (Table 9). Statistics are not available for access to higher education except for Iran, Algeria, Jordan, Syria and Egypt where some numbers are shown in maths and engineering.

Although these increases in public spending are much welcomed, they are still regarded as inadequate by the internal critics in many of the Arab states. Said Aburish (*The House of Saud*) cites a petition in 1994 from 124 ulemas to King Fahd of Saudi Arabia which called for the curtailment of arms purchases and reallocation of the money to health care and education. Considering the relatively high levels of national income in many of the Arab States, welfare expenditure cannot be regarded as particularly impressive, and the ulemas were right in seeing that the reason was not far to seek. Quite apart from what are reported by Said Aburish to be the 'princely allocations' in Saudi Arabia, which are allegedly deducted from the oil revenues before the national income is declared – and these may amount to between $4 billions and $7 billions in a year – there is the vast scale of military expenditure. This is not limited to Saudi Arabia. *Jane's Defence Weekly* in 1997 (February 6th and 19th) and again in 2002 (January 9th) reported complaints from Committee inquiries in the Kuwait Parliament of corruption and wasteful spending on arms procurement.

The Scale of Military Expenditure and Arms Purchases

High levels of public expenditure on the military in all the Arab States show where much of the oil wealth has been going. Although this military expenditure had been reduced to a half by 2001 from even higher figures ten years earlier, it still amounted to over 6% (11% in Saudi Arabia) of the national income (GDP) of the Arab States, to be compared with about 2% in Europe and 3% in the United States. In Iraq in the 1980s military spending was taking over 40% of GDP, more even than in Israel. In the 1980s in some countries, Saudi Arabia, Oman and Kuwait, such expenditure was at 20% of GDP (Table 10). A part of this military expenditure went to maintaining large armed forces. Turkey, Iran, Syria and Egypt had the largest numbers – from 300,000 to half a million under arms. In all the Arab States the total numbers exceeded two million (Table 10). But far the greater sums were spent on imported weapons, amounting in the years from 1995 to 2002 to over $100 billions on some estimates. The scale of such military spending by Saudi Arabia was particularly large, although recently reduced (Table 11)

13

Estimating figures for international arms sales presents even greater difficulties than those for oil sales. Only very limited figures are available in most of the official statistical publications. Arms are not normally distinguished in a country's lists of imports and exports and may appear with other motor vehicles, transport equipment or metal products and other components, or may appear under 'services' and not 'goods'. Or they may simply not be recorded in official statistics at all. Some information can be obtained either from the annual studies issued by the International Institute for Strategic Studies (IISS) or from the Stockholm International Peace Research Institute (SIPRI). The arms transfers detailed each year in the *SIPRI Yearbook* are deliberately referred to as 'transfers' and not as 'trade', because the arms are sometimes not sold, but may be given as aid or as part of some barter, leasing or other arrangement such as literally 'arms for oil'. Government to government transfers are generally not recorded.

The SIPRI arms transfer figures are the most detailed, but the totals year by year are shown in the form of what SIPRI calls 'trend indicator values' over five years, the latest being 1998-2002. These are based on the value of specific deliveries over the years, not of sales or signed agreements. SIPRI claims that they thus represent the actual military resources being transferred. All figures are related to the price of each item in a base year to show constant prices (1990 the most recent year). Partly because of rising prices over the years, SIPRI gives a much lower average annual figure of arms imports than the IISS which takes 2002 as the base year: an annual average of $18.5 billion for all arms deliveries over the years 1998-2002, instead of $32.7 billion over 1999-2002. And it is the SIPRI form of calculation giving the lower figure which the UN Yearbooks employ (Table 14).

There must be some other reason beside rising prices, however, for the size of this difference in the two sets of figures. The IISS estimates for deliveries of arms to Saudi Arabia are quite remarkably larger than the SIPRI estimates, showing these deliveries to be over half of the total to the 'Middle East' and North Africa in one case and no more than a fifth in the other (Table 14). Remembering that the SIPRI estimates are based on the actual cost of items of equipment in the base year and on further orders of similar equipment in succeeding years, while the IISS estimates are based on recorded expenditures year by year, there may be expenditures on associated training, agency fees, commissions, etc. included in the latter which do not appear in the former. The implications of such divergences will be explored later.

The fact is that information on arms sales is subject to many forms of state secrecy and security restrictions. These have been much increased by the attempts of the United Nations to impose embargoes on supply to warring parties. This was most prevalent in the case of Israel and of the Iraq-Iran war, when major supplying countries – the United States, United Kingdom, France and Russia – sought to evade the embargoes by routing supplies to non-belligerent countries, Egypt and other Gulf states or further afield, for re-export, or by subcontracting suppliers in Yugoslavia, Spain, Thailand, Eastern Europe and elsewhere. This is not to be blamed on 'wicked arms dealers', but on the secret policies of governments.

Wherever estimates of arms deliveries are made public, the 'Middle East' is shown to be the recipient of at least a quarter of the total world-wide, and in some years this rises to a third, with Turkey, Saudi Arabia, Kuwait and the United Arab Emirates at the top of the most recently published list in 2002, along with China, India and Israel (Table 14). These proportions must be compared with populations in the 'Middle East' countries comprising seven per cent of the world total (Table 1), and such estimates must understate the reality, because it has mainly been in the 'Middle East' that wars have raged, involving the attempted evasion of UN embargoes.

In considering the importance of all military expenditure for 'Middle East' countries, we have seen the proportion of national income (GDP) assigned to the military (Table 10), but there are other more telling ways of expressing this. One is to look at such expenditure on defence as so much *per capita*. In this respect, not only Israel but also Saudi Arabia, Qatar, the United Arab Emirates, Oman and Jordan spent more per head of population on arms in both 1985 and 2002 even than the United States (Table 11). Another way is to look at the proportion of a country's total trade consisting of arms. Saudi Arabia is also in the lead here with 17%, far above the next two, Kuwait and Egypt with around 7%-8% each (Table 13). Even the

15

average for all 'Middle East' countries comes to 10% and may be compared with 2% of US exports consisting of arms and 4% of UK or Russian exports. The importance of arms in the trade of 'Middle East' countries is clear, but not so clear at first sight in the trade of the West.

While arms exports in the 1990s made up only 1% of all world exports, the proportions were only somewhat larger for the main industrialised countries of the West. They had previously been considerably higher in the 1980s, although never more than a little above the overall proportion of expenditure on defence in these economies (Tables 11 and 13). We cannot say from this evidence that the arms trade was the driving force of the Western economies, though we have to notice the scale of the arms sales of some of the largest companies. In the year 2000 three US companies – Lockheed Martin, Boeing and Raytheon – had sales of arms in excess of $14 billions each and four others had arms sales of $4-$5 billions. From the United Kingdom, BAe Systems had annual sales of over $13 billion having benefited from the privatisation of the Royal Ordnance. France's largest arms companies – Dassault, EADS and Thales – each had sales of over $4 billions (Table 16). These companies, though large, barely scrape into the world's top 500, and we shall have to consider what other influence on governments they may have. In doing this, a key question must concern the reasons for the concentration of this arms trade on the 'Middle East' – a third of all world arms sales despite the small size of these countries – in all having populations not even of 400 million including North Africa out of a world population of over 6000 million.

Why the Huge Scale of Arms for Oil?

The large scale of the world's arms trade seems to come down to the particularly large deliveries of arms to the 'Middle East'. We have, therefore, two problems to consider: why military expenditure in the 'Middle East' countries is so high? and why arms exports of the industrial countries to the 'Middle East' are so large? Both problems might be answered by instancing the volatility of the government regimes and the interest in security of oil supplies. But it will be argued in this paper that the direction of causation is not what one might suppose from the relatively small weight of arms in the industrial countries' total trade compared with that in the 'Middle East' countries' trade. It is not, in fact, so much the case that military expenditure of the 'Middle East' countries requires large imports of arms from the main industrial countries, but the other way round: that the exports of arms from the industrial countries requires large military budgets in the 'Middle East' countries. How does this come about?

There might be four possible explanations for heavy arms expenditure by Arab states – fear of Israel, fear of Communism, fear of their neighbours, fear of their own people by the ruling élites. Each of these fears undoubtedly has some substance. Fear of Israel could be the largest factor because of the need seen by some Arab states for private guarantees of security in exchange for arms purchases, but it has not resulted in any attempt among the Arab states to coordinate their military capabilities. They gave little help to Nasser in his wars with Israel, being rather more afraid of Nasser's socialism and nationalism. Apart from Iraq they have made no sustained attempt to build a nuclear response to Israel's nuclear power. Only Iran, a non-Arab Islamic state, has sought to do that. The Arab states have in the main relied on the assumption of American protection for their defence. Fear of Communism in the form of Soviet influence did not deter either Nasser or Saddam Hussein from buying arms from the East as well as the West. Fear of their neighbours appears to be an abiding preoccupation especially among the Saudis, for whom the growth of Arab nationalism is of particular concern. With the collapse of the Soviet Union, this fear of nationalism combines with fear of socialists and other dissident groups among their own peoples to provide a strong reason for the Arab élites to maintain and retain large armies and police forces. But this cannot explain the type of armament imported – the most advanced military aircraft and air defence systems that the West has on offer (Table 17).

The alternative explanation for the scale of the oil for arms trade between the West and the 'Middle East' is that this trade is deliberately encouraged by Western suppliers for some reasons of their own. A powerful argument in favour of this view is that several of the recipient countries, including Saudi Arabia, are said to be incapable of making use of such quantities and types of arms. The implication is that they have been forced, or at least enticed, to buy them. The

International Institute for Strategic Studies (IISS) in its 1989 report commented that 'The Saudi army is not growing in proportion with the hardware at its disposal', and has regularly expressed doubts about the capacity of the Saudi air force and navy to make use of the most technologically advanced US aerial defence systems and French torpedo boats. A report in the *Washington Star* (18.05.87) claimed that the Saudis could only operate one out of five of the AWACS planes they had leased from the US. In July of 1988 there were articles in *The Economist* (18.07.88) and the *Financial Times* (12.07.88), claiming that the Saudis lacked the personnel for the arms they were purchasing.

The waste of resources is staggering, and the assumption must be that corruption is endemic – a point which will be discussed at length later. During the first Gulf War, Robert Fisk of *The Independent* newspaper reported that he had found a dump of hundreds of trucks, which the Saudis had forgotten all about. When the US contractor Bechtel built the King Fahd International Airport for $3.4 billion, one disappointed contractor said that he could build five airports for the price of that one. Costs per soldier, according to IISS figures, amounted in 1982 to $470,000 in Saudi Arabia compared with $103,000 in the US and less than half that in the United Kingdom, Germany or France. To take one example, according to Said Aburish (*The House of Saud*, p.199) catering contracts to feed US troops during the Gulf War varied from $70 to $133 a day, compared with $30 a day to feed French soldiers.

The reasons for the arms excess could be several. First, the governments supplying arms may believe that arms sales are the only way of obtaining the oil that they need. This can hardly have been the rationale for the particularly high level of arms sales to the 'Middle East' in the decades of the 1980s and 1990s, when the main arms deliveries were coming from the United States, United Kingdom and Russia, which all had their own considerable supplies of oil (Table 3). Second, the Western arms suppliers may have believed that in arms they had a particularly high competitive edge against other buyers of oil, Japan for example. This can't be true either, because, at least in the case of the United States and United Kingdom, arms exports were being heavily subsidised by the governments of the exporting countries (Table 18), concerning which subsidies more will be said

18

later. A third possible reason has to be considered – that the main arms manufacturing companies in the West alongside of the big oil companies had some special relationship with their governments that encouraged the arms for oil trade.

Special relationships may exist between the governments and the big armament and oil companies based in their countries because of the governments' concern for their people's security in the event of military or economic threats to their livelihood. Control over supplies of oil and of arms will be an important national policy. Having an arms industry or an oil industry on a scale that is competitive world-wide may depend on having wider markets than the home market. The special relationship will then consist of state subsidies, especially for exports and for advanced industrial research, which underpin the country's competitive position in high technology. It is not for nothing that President Eisenhower referred to the power in the United States of what he called the 'Military-Industrial Complex'.

This all has to be understood in the framework of the new stage in the development of capitalism after about the early 1970s. This has been described by David Harvey in his *New Imperialism* (OUP 2003) as 'accumulation by dispossession' which, he proposes, has replaced 'accumulation by extended re-production'. Capital is accumulated not only by profit making, by merger and take-over, including asset stripping, but by dispossession – the privatising of public assets, suppression of small businesses, buying up of small farms by agribusiness, debt peonage, raiding of pension funds, corporate fraud, opening up of closed economies and, of great importance in the arms trade, by state subsidies and state guaranteed credit. This is the new wave of 'enclosing the commons' upon which capitalism was founded and is now pursued in the name of neo-liberalism.

Oil companies and arms manufacturing companies tend, moreover, to be among the largest companies in those countries which have such companies based in their lands. The size of the oil companies is well known, but the arms manufacturers are almost equally large. Their influence is likely to be correspondingly big. The scale of the subsidies provided by governments to arms sales from the United States shown in a Table from the *SIPRI Yearbook 2003* (page 544) is astonishing. In the years 1992-2001 a total of over $40 billions

($4 billions a year) was administered by the Department of Defence, including grants of $35 billions, loans of $3.5 billions, plus $2billions from the Military Assistance Programme and $1billion from US Export-Import Bank guarantees for financing military export programmes and loans (Table 18). And what is shown here does not cover everything. Because of limits placed on these subsidies in the United States, some $5 billions of arms deliveries in the 1980s were actually made under the guarantee of the (Agricultural) Commodity Credit Corporation. In the United Kingdom subsidies costing the taxpayer many billions of £s have been revealed in several studies by the Campaign Against the Arms Trade (CAAT, 2003/2004) and by Stephen Martin, *Gunrunners' Gold* (1995), Paul Ingram and Ian Davis, *The Subsidy Trap* (2001) and Paul Ingram and Roy Isbister, *Escaping the Subsidy Trap*. Secret subsidies became particularly important when arms exports were banned during the Iran-Iraq war. According to Gerald James in his book *In the Public Interest*, US helicopters fitted with air-to-ground missiles were labelled at the 1989 Baghdad Arms Fair as 'For Agricultural Use'.

In the United Kingdom in the past, 20% of the state funds available under the provisions of the government Export Credit Guarantee Department (ECGD) were reserved for arms exports, but it is generally assumed that under Mrs. Thatcher's regime the proportion was very much larger. Under New Labour it has amounted, according to the Campaign Against the Arms Trade, to 30% to 35%. Gerald James records (*In the Public Interest*, pp.109-111) the scandal of the disappearance, in 1984, from the ECGD accounts of £83 millions, on which James comments that this sum bears a remarkable similarity to a sum paid from the Jordanian arms deal package, but not accounted for in actual contracts. Robert Sheldon, chairman of the Public Accounts Committee of the House of Commons, expressed the opinion that the total unaccounted for might be much larger, but he allegedly refused James' request to investigate the matter further. In the illicit supply of arms to both sides in the Iran-Iraq war, it appears from a memorandum circulated in February 2004 to Members of the European Parliament by Sir William Jaffrey, one of the Lloyds Insurance 'names' who lost their money in the 1990s, that a sum of more than £73 billions (yes, billions at 1990 prices) was provided by the ECGD, on which Iran and Iraq both defaulted. Robin Cook in his book *The Point of Departure* (Simon & Schuster 2003) admits to the UK Treasury finding £1 billion for contracts on which Saddam Hussein had reneged.

It is simply not known what sums were involved in financing the Al Yamamah deal. A report was made by the Auditor General, Sir John Bourn, who was appointed by Mrs Thatcher in 1988, and was previously a senior Ministry of Defence procurement official. But this report remains secret. Even the Public Accounts Committee of the House of Common was not informed, according to what is recorded in Gerald James' book (*In the Public Interest*, pp. 61ff.). At the same time, James quotes Lord Young, one of Mrs Thatcher's cabinet ministers, reporting to the Public Accounts Committee in 1988 that the £120 billions from North Sea oil revenues (surprisingly, just the same sum as the total value of the Yamamah arms deal), well invested, would bring in $5billions a year dividends from overseas. These investments, Lord Young averred, would make up for any disappearance of British manufacturing industry. In fact in 1990 British overseas assets suddenly dropped from £64.6 billions to £-0.4 billion, just when ECGD was most hard pressed to settle overseas credits. It seems that these huge losses left a black hole in the British Treasury finances, which the sell off of state assets by British governments from Mrs Thatcher onwards has had to make good.

The extent of the influence of the big oil and arms manufacturing companies is not widely known. This is partly because national security is involved and a veil of secrecy surrounds much of their relations with governments. Those who have been involved as arms suppliers, like Gerald James, know that absolutely no arms exports are permissible from either the United States or United Kingdom without government sanction. Most are sold to the state. From time to time the veil of secrecy is rent apart and some uncomfortable truths for government are revealed. This is the case when the finances of the two big political parties in the United States are divulged and the oil and arms business connections of Presidential candidates like George Bush, father and son, and of vice-presidents like Dick Cheney, who was president before joining the Bush government of Haliburton Corporation, a major military construction contractor. These companies support powerful lobbies such as the 'US Committee to Expand Nato', presided over by the Chief Executive of Lockheed Martin, the largest arms producing company in the United States.

The House of Bush – House of Saud Relationship

In the years of Ronald Reagan's Presidency of the United States, a new relationship was established between the US and Saudi Arabia which was to have dire consequences for the peoples of both countries. This was the strong personal relationship between the Bush family and the Saudi royals, most especially Prince Bandar, which is described in detail in Craig Unger's book *House of Bush: House of Saud* (Gibson Square Books, 2004). This began in 1981, with the US agreement to a $5.5 billion package of AWACS planes and all the associated technology to go to Saudi Arabia. This was the start of a decade of Saudi orders for US arms, airfields and ports valued at some $200 billion, in which a major beneficiary was the Saudi Bin Laden construction group, whose immense wealth Osama Bin Laden was to inherit.

Unger's story runs as follows: George HW Bush, who had settled in Houston, Texas, was Reagan's vice president. Prince Bandar, with a Master's degree at Johns Hopkins University, was leader of the lobby in 1980 to persuade the President and Senate to agree to this huge arms package against powerful resistance from the Israeli lobby. Bandar won and was made Saudi ambassador to the USA, and remained at that post for twenty years. Bush senior and Bandar became close friends, both enjoying the cowboy sports of the Texas ranches, horse riding, hunting and shooting. Saudi money had been flowing into Texas since the oil price hikes of the 1970s because of the opportunities opening up for investment in this then still undeveloped part of the United States, where oil finds were beginning to attract new money. Some of this Saudi money was being directed by the Saudi Bank of Credit and Commerce International (BCCI), which came to a very sticky end, allegedly, as the bank of choice for illegal arms sales to Iraq as well as for other covert CIA operations.

Much of the Saudi money, however, was directed by the US Carlyle Group of defence contractors with assets of over $16 billion. Carlyle Investment was founded by David Rubinstein and became what Craig Unger describes as 'home to George HW Bush, James Baker (Bush senior's campaign manager and Reagan's chief of staff), Frank Carlucci (Reagan's Defense Secretary), Richard Darman (head of the Office of Management and Budget), John Major (one-time UK Prime Minister) and other powerful figures from the Reagan-Bush era.' 'An element in Carlyle's ascendancy,' Unger goes on to explain, 'has been its relationships with the Saudi royals, the Bin Ladens and the Mahfouz family both as investors and clients for defence contractors owned by Carlyle.' Many of these Carlyle associates had a background in Intelligence – George HW Bush had been CIA director for the years 1976-80 – and together with Bill Casey, who followed Bush at the CIA, the group became what Unger describes as a 'cabal' directing US foreign policy. Casey is reported by Unger to have claimed that he depended on the Pakistani founder of BCCI, Hasan Abedi, for all the secrets of Middle East leaders.

The new relationship of the US with Saudi Arabia has to be understood in the light of events in the 'Middle East' that preceded the signing of the AWACS agreement in 1981. When the Shah of Iran, the USA's main ally among leaders of oil states, had been overthrown in 1979 by a Muslim fundamentalist coup, hostages were taken in the US embassy in Tehran and President Carter's delayed ability to get their release led to his election defeat by Reagan. The US alliance was switched to Iraq, and Saddam Hussein was encouraged to invade Iran. Despite UN decisions that all states should remain neutral, the United States and United Kingdom supplied Iraq with arms including biological and other weapons of mass destruction, which Saddam used to defeat the Iranians after seven years of war.

In the same year of 1979 two other events took place in the 'Middle East' which encouraged closer ties between the Saudis and the United States, but which greatly complicated the relationship. The first was the decision of Zbigniew Brzezinski, President Carter's national security adviser, to entice the Soviet Union into war in Afghanistan, believing, as it turned out correctly, that this would be the Soviets' Vietnam. Many hundreds of Muslims from Saudi Arabia and elsewhere were encouraged to join the *mujahadeen* guerrilla forces in Afghanistan to eject the 'infidel' Soviet invading army. These came under the leadership of Osama Bin Laden, whose brother Mahrous is said possibly to have been one of the thousand members of the Muslim

Brotherhood who had invaded Mecca and seized control of the Grand Mosque in protest against the profligacy of the Saudi royal regime. This was the second event of 1979 which led the Saudis to look for arms from the United States, but which at the same time created the terrorist forces that were to strike back at the US itself in 2001.

When the Soviet army withdrew from Afghanistan in 1988, the Taliban took power, with Osama Bin Laden's Al Qaeda establishing a base for world-wide terrorist activity. Meanwhile, Saddam Hussein having defeated Iran, took one step too far for his American masters and, in 1990, invaded Kuwait. The United States new relationship of friendship with the Saudis enabled the Bush government to establish a main base in Saudi Arabia for its war against the invading Iraqi armies, although Osama bin Laden warned against it. The Iraqi armies were driven back or destroyed but US forces did not go on to Baghdad. The United Nations imposed sanctions on the Iraqi regime including the requirement that all weapons of mass destruction should be eliminated. The US forces remained in Saudi Arabia. Their continued presence deeply offended Osama bin Laden and all Muslim fundamentalists. That the infidel should be allowed to set foot in the holy land was unacceptable. Acts of terrorism against the US at the World Trade Centre in 1993, and inside Saudi Arabia itself in 1995, were confirmed as actions of Al Qaeda with Osama bin Laden's declaration of *jihad* against the United States in 1996. This culminated in the bombing of US embassies in Dar es Salaam and Nairobi in 1998, and of the USS *Cole* in Aden in 2000. President Clinton responded to the embassy bombings by the United States bombing Al Qaeda camps in Afghanistan and a pharmaceutical company in Sudan. Before getting ensnared in the Monica Lewinsky affair, he did strengthen the position of Richard Clarke as head of the National Security Council's Coordinating sub-group into that of a kind of counter-terrorism czar.

Richard Clarke's comments on the Saudis quoted by Unger are illuminating:

'There's a realisation that we have to work with the government we've got in Saudi Arabia. The alternative could be far worse. The most likely replacement to the House of Saud is likely to be more hostile, in fact extremely hostile, to the US.' But Unger argues that 'if the House of Saud were a genuine ally, the Bush administration could have pressurised it

24

about the Saudi role in terrorism, aggressively gone after Al Qaeda after the USS *Cole* bombing and still maintained a productive alliance. But that didn't happen.'

Richard Clarke, according to Unger, seems never to have ceased demanding action against what he believed was a real threat from Al Qaeda. As late as September 4th, seven days before 9/11, Clarke was insisting that a strike within the United States was imminent. Unger quotes Charles Lewis, head of the Centre for Public Integrity in Washington, asking the crucial question: 'When it comes to war on terror, a lot of people have to be wondering why we are concerned about some countries and not others. Why does Saudi Arabia get the pass?'

Unger makes it quite clear that he thinks the House of Bush-House of Saud personal relationship was the problem. He quotes Charles Lewis again, addressing the particular issue of Bush-Saud ties within the Carlyle Group. 'You would be less inclined to do anything forceful or dynamic if you are tied in with them financially'. To this Unger comments, 'It is a factor that more than $1.4 billion has made its way from the House of Saud to individuals and entities tied to the House of Bush.' What happened after 9/11 only goes to support Unger's argument.

Two otherwise inexplicable events followed. On September 12 all planes were grounded throughout the United States. But Prince Bandar was able to organise flights across the United States on that day for the later evacuation by plane from the country of 140 members of the Saud and Bin Laden families without interrogation and with the evident authority or at least acquiescence of President Bush. Somewhat later, when *Newsweek* cited Bandar's wife Princess Haifa as a possible source of funding for two of the 9/11 hi-jackers, Unger reports that Laura Bush and ex-President George HW Bush called to console her. Further evidence of Bush junior's softness for the Saudis and bin Ladens comes from the ease with which he was converted by the Neocons in his cabinet to elide chasing Al Qaeda with displacing Saddam Hussein. George Bush junior had been brought up inside the family friendship with Saudis and enjoyed even more than his father the cowboy life of Texas, which so attracted Prince Bandar. But he took into his cabinet from the Republican Party the leading members of the Project for the New American Century (PNAC). A part of this project was to Rebuild America's Defences, including a forceful US military presence in the Middle East.

Donald Rumsfeld, who had been a special envoy of Bush senior, twice visiting Saddam Hussein to reassure him of US support during the Iraq-Iran war, was a PNAC signatory, and became Bush junior's Defence Secretary. But he had turned against Saddam Hussein, and believed that control of Middle East oil reserves demanded regime change in Iraq, and a position for the United States in the Middle East that was not dependent on Saudi goodwill. Vice President Dick Cheney and Deputy Secretary of Defence Paul Wolfowitz and other leading members of the new Bush government were also PNAC signatories. So the new Bush team was very different from the old Bush cabal with its close Saudi relationship.

Richard Clarke resigned his counter-terrorist post in February 2003, just before the invasion of Iraq, commenting to journalists, 'You know that great feeling you get when you stop banging your head against a brick wall.' The invasion of Iraq was launched, however, only after allowing for the sensitivities of the Saudis, by moving the US forces to launch the invasion out of Saudi Arabia and into Kuwait and other Gulf states.

The British Intelligence Cabal

It now becomes somewhat easier to understand the evident determination of the intelligence services in both the United States and the United Kingdom to distance themselves after the event from the decision of the US followed by the UK government to go to war in Iraq. British government relations with the 'Middle East' are not so easily discovered, as are those of the United States. But incidents such as that of the supergun being manufactured in Sheffield for Iraq, which became the subject of the Scott Inquiry, do occasionally reveal what is happening behind the scenes – never the whole story. The Scott Inquiry ran to several thousand pages, covering five years of collected evidence with 20,000 source documents, but Lord Scott never thought to examine the financial arrangements involved. The story is told in a thesis submitted for examination by Gerald James' son, Alexander. In a private communication from Gerald James it is claimed that Lord Scott never called as a witness Sir John Cuckney, who was chairman or vice-chairman of the main companies – Midland Bank International Trade Services, Westland Helicopters, TI Group, Matrix Churchill and 3i – all allegedly involved in the supergun affair. Sir John Bourn, moreover, at the age of 70, remained the UK Auditor General, holding all his secrets close to his chest.

The whole story of the arms deals made by Mrs Thatcher, first with Jordan and Malaysia and then in the infamous Yamamah contracts with Saudi Arabia, perfectly illustrates the nature of the arms for oil relationship. Mrs Thatcher was not only able to revive some declining British industry but out of the commissions, it is alleged, to finance the Conservative Party. To understand how this could happen, it is necessary to follow the complex arrangements of intermediaries that are involved in reaching an arms deal with any oil producing state, but most particularly in dealing with the Saudis. The first and absolute essential is a government contact (not a contract; that comes much later). In Saudi Arabia the contact is a Saudi prince, according to Said Aburish (*The House of Saud*, Bloomsbury 1994, pp. 182 ff.). There are some 7000 princes, but those nearest to the king are the ones that matter. A sales agent must start with an intermediary and work up to the 'intermedler' or 'skimmer', a royal member who can decide who gets the deal. On top of all this there are fees to be paid for consultants, some of whom may be very necessary to any deal. Others have been found to be non-existent, but still appear in the accounts.

At each stage, of course, there is a payment to be made, politely called a 'commission', less politely ' a bribe'. The most famous super-intermediary in Saudi Arabia, according to Aburish, was alleged to be Adnan Khashoggi, a close friend of King Fahd and of the Minister for Defence Procurement, Prince Sultan. Khashoggi allegedly got business for such major companies as Lockheed, Rolls Royce and Grumman. Jonathan Aitken, as UK Minister for Defence, became famous, and lost his job and went to jail, when he was found to have lied in court about having his Paris hotel bill paid for by King Fahd's son Muhammad. Mrs Thatcher's son Mark was alleged to have had similar connections. Prince Sultan bin Abdel Aziz, as not only Minister of Defence Procurement but officially number two in line to the throne, is said by Aburish to be the leading example of an 'intermedler' and certainly one of the world's wealthiest men.

The sums involved in commissions are staggering. Twenty five per cent of the contract price is not unheard of. If purchases of arms in any year exceed some $10 billion, and that is not unusual – the Yamamah contracts alone amounted to $150 billion over 20 years, according to BAe chairman John Cahill – then bribes of some $4-$5 billion a year might be involved. Gerald James quoted to this effect documents which were revealed in connection with the case of a certain Lieutenant Colonel in the United States, who claimed compensation for wrongful dismissal against United Technologies Corporation, when he blew the whistle over an alleged bribery scheme involving Westland Helicopters. Work was to be charged for as if done by a non-existent Saudi company, work which was actually being done elsewhere. The sums involved in obtaining an alternative engine 'no-deal' in favour of BAe and Rolls Royce amounted to $4 billion. Neither BAe nor Rolls Royce could find that sort of money. So, Gerald James concluded that it must have been the British Government putting up taxpayers' money, as in the ECGD financing of the illegal Iran-Iraq sales.

The financing of arms deals both in the United States and the United Kingdom involves, as we have seen, a remarkable amount of subsidy. The question being raised in a number of British journals in the early 1990s, in particular the *Sunday Times* and the *Business Age* magazine, was how much of British taxpayers' money was finding its way into the coffers of the Conservative Party. Sums of the order of £100 million were flowing in as unaccounted donations during the years 1985 to 1993, the very years of Mrs Thatcher's big arms deals. *Business Age* in February and May of 1993 claimed to have discovered slush funds at one time totalling £200 million. The magazine mused that there might be a connection with Mrs Thatcher's son Mark who was alleged to be involved in some of his mother's arms deals, and famously commented when later collecting funds for the Thatcher Foundation that 'it is pay-up for mumsie time'. Gerald James points out that one of the known donations came from a Chinese billionaire of the Hong Kong Shanghai Bank, who was also a member of the Chinese Parliament and western representative of the Chinese arms firm NORINCO. The Hong Kong Shanghai Bank (HSBC) in 1988 acquired the Midland Bank, the bank whose Midland International Trade Services was at the very heart of the British arms industry in the 1980s.

Who are the real actors in East-West economic relations?

In order to describe the nature of economic relations between the 'Middle East' countries and the West, it has been necessary to refer to governments both in the West and in the East, to the giant oil companies which extract, transport, refine, process and retail the oil, to the transnational corporations which manufacture the armaments, to the banks which finance the whole business, and to the various intermediaries involved. It has been suggested that those personally involved in the business may have their own agenda that are concealed behind the veil of national security. This sounds very much like a conspiracy theory, but there is a host of evidence for the view, although not everything can be generalised from the particular relations of the Bush family or of Mrs Thatcher with the Saudi royal family.

Many questions have been asked in recent years, particularly in the United Kingdom, by the Campaign Against the Arms Trade about continuing New Labour subsidies for military exports, which were still at the end of the Twentieth Century estimated to amount to up to some £1 billion per annum (Emma Mayhew, 'A Dead Giveaway: A Critical Analysis of New Labour's Support for Arms Exports', a paper delivered to the annual British International Studies Association Conference at the London School of Economics, 2002, to be published in the *Journal of Contemporaty Security Studies*, March 2005). Apart from arguing that New Labour was only continuing contracts that the previous Tory Government had signed, the main rationale was said to be the protection of the employment of the large numbers of workers involved in the arms industry. It was argued that under New Labour numbers employed on arms exports had been reduced from some 200,000 to nearer 70,000, of whom about half were directly dependent on the arms trade. Assuming that 50,000 were then benefiting from the £1billion annual subsidy, this amounted to about £20,000 a head. Ten years earlier the Export Credit Guarantees funded by the UK tax payer were amounting to tens of billions of £s, not £1 billion, for a workforce no more than perhaps three times the size.

It is evident that the subsidies were far in excess of what was needed to maintain employment. Emma Mayhew in the paper referred to had in any case been able to demonstrate that workers laid off by reduced arms exports had already been absorbed without difficulty into other employment, and that more could be found re-employment if necessary. Another argument deployed in favour of subsidising arms exports was that the industry needed subsidising so that the country had its own arms supplies and was not dependent on others. This has been answered in a study by the Campaign Against the Arms Trade (January 2005). Some 40% of British arms production already consisted of imports from other countries, and many of these were the result of inter-national internal company agreements which could not readily be terminated. Antonia Feuchtwanger in a study published in 2004 ('The Best Kit' published by The Policy Exchange) argues that, because Britain's military industry is protected from competition, it fails to provide Britain's armed forces with the best equipment, that the government should stop trying to prop up 'UK Defence Plc', allow contractors to be taken over by foreign ones and buy more equipment from abroad.

If there was no national advantage in maintaining large British arms exports, it was likely that the same arguments applied in the United States or in France or Russia, where arms exports remain a similarly small part of total business. What, then, is the explanation for the massive arms export subsidies, the deals with Saudi Arabia in particular? Reference has already been made to Gerald James' book *In the Public Interest*. This book is not primarily about relations with Saudi Arabia, but more generally about US and UK arms deals with 'Middle East' states during the Iran-Iraq war. United Nations resolutions had required arms supplying states to desist from supplying arms to either side. The burden of James' book is that the embargo was broken by the United States and the United Kingdom and other governments through two main means. First, the belief was encouraged that exported goods had dual use, both civil and military. In this connection Alan Clark, a British Government Defence

Minister, first used the phrase that 'one can be economical with the actualité'. Secondly, goods were ordered for and dispatched to a non-belligerent country, from which they could then be forwarded to one or other of the belligerents.

The central argument of the book by Gerald James, who was himself founder of a medium sized arms manufacturing company with 4,000 employees in factories in the United Kingdom, United States and Belgium, was that this practice was authorised by the Ministry of Defence in the UK and by the US Defense Department as the result of quite special relations inside governments between the defence departments and certain leading arms supplying companies and their chief shareholders. Small and medium sized arms suppliers were not informed of what was happening to their exports. When James found out and blew the whistle, he was ousted from the chairmanship of his company by a government organised coup, had all his papers seized by the police, lost his licence to do business and was unable to obtain compensation. I have met the man and I believe him.

The relevance of Gerald James' story to a study of East-West economic relations is that in the United Kingdom and the United States, and probably in France also, there is what James called a 'cabal' which determines these relations, a 'cabal' which is based on the secret services in each of these Western countries which have their 'spooks' among government ministers, civil servants in defence departments, ambassadors, directors of certain banks, and of the big oil and arms manufacturing companies. Gerald James is able to name names in the United Kingdom of the particular ministers, defence department and foreign service officials, bank directors and company chairmen all with secret service backgrounds. Most Prime Ministers in the UK and some US Presidents are unaware of the existence of these 'cabals', or may even be the object of their attentions, as in the case of Harold Wilson and Bill Clinton. Mrs Thatcher, with her husband's oil business connections, like those of the Bush family, kept herself closely informed of secret service activities, which allegedly gave her and her son much insight and influence in the arms for oil business.

Nor has this close relationship between UK Governments and the arms industry ended with the advent of New Labour. Tony Blair came to office with the promise that government in future would pursue policies, especially in foreign affairs, which would be strictly ethical after the previous Tory administration had collapsed in a cloud of rumoured sleaze. Those who believed Mr Blair were in for a sad disappointment.

The Campaign Against the Arms Trade in its study 'Who Calls the Shots?' (published in February 2005) concludes:

> 'Military industry has attracted a seamless progression of top ex-MoD and armed forces staff and has offered numerous members of its own staff to the MoD as secondees, blurring the boundaries between industry and government. It is an industry that is extremely heavily represented in a network of bodies advising the government on military policy...It is an industry in which a number of the most senior former executives now enjoy influential government positions as Labour peers and/or have links to some of the highest echelons of the government. It is an industry that has contributed largely undisclosed amounts of cash to the Labour Party and to one of its major projects, the Dome. The contributions have formed part of a wider Labour Party trend towards financial dependence on companies and wealthy individuals which, in turn, provides the Party with a material incentive to pursue business-friendly policies.'

To take but two examples of the evidence: first, Lord Hollick, a banker who was a director of BAe from 1992-1997, became Special Advisor to successive British ministers at the Department of Trade and Industry. It was alleged in *The Observer* that he contributed to the watering down of Foreign Secretary Robin Cook's attempts to regulate arms exports. Second, Richard Evans, chair of BAe in the 1990s, was at least until January 2003 the chair of the National Defence Industries Council, the Defence Industries Council and the Aerospace Innovation and Growth Team. Richard Evans was called to answer questions before a House of Commons Select Committee as the result of the whistle blowing by ex-staff of BAe concerning bribes made to Saudi royals, reported in a BBC 2 programme 'Bribing for Britain', broadcast on 5th October 2004. The organisations which Richard Evans chaired are all part of the main government/industry Task Force for advising the government and initiating research and technology innovation in the arms industry. They involve the chair and/or the chief executive officer of all the main arms producing companies, along with ministers and civil servants from the MoD, often themselves seconded from industry, but no representatives of trade unions or other political associations, so far as is known.

Unilateralist US Oil and Arms Policy World-wide

It is only such insider dealings that can explain the extraordinary history of the arms for oil relationship that developed between the West and the 'Middle East' regimes throughout the last century, and that continues to determine these East-West relations today. There is the continued support for a seemingly utterly corrupt regime in Saudi Arabia and similar regimes nearby, the change in attitude to Saddam Hussein from the most favoured and best armed leader in the 'Middle East' to the 'axis of evil' requiring two major wars to unseat, the supplying of arms to both sides of the Iraq–Iran war against UN embargoes, the arming over many years of brutal regimes not only in Iraq but also in Turkey, the recruitment of Osama Bin Laden to mobilise *mujahadeen* to harass the Soviet forces in Afghanistan, the same as leader of Al Qaeda, who became enemy no.1 after 11/9/2001, the propping up of separate states in Jordan, Lebanon and Syria and the Gulf states of Bahrein, Oman and the United Arab Emirates, and finally the new unlilateralist posture of US oil policy.

All this can only be understood by holding in one's mind the machinations of spymasters and military commanders, wholly unaccountable to any democratic process, who see that the best service to their employers lies in playing a permanent game of divide and conquer, whatever the long term consequences for the peoples involved, and whatever the response of the international community as expressed through the United Nations. This is not to say that the intelligence supplied by the secret services to their masters is irresponsible. Such intelligence is often, as was revealed after the inquiries into the information supplied to the British and US governments in advance of the second Iraq war, a great deal more cautious than the governments wished to make use of. But this was because of the unilateralist posture that President Bush and his Neocon advisors had adopted.

It has to be understood that 'intelligence' is not only concerned with military and political affairs, but very particularly with economic relations. The British Joint Intelligence Committee, which became the object of much interest in the inquiries in 2003-4 into the 'intelligence' on the basis of which Mr Blair took Britain into the war in Iraq, revealed its priorities as early as a note in 1969. According to Mark Curtis in his book *Web of Deceit: Britain's Real Role in the World* (Vintage 2003), this note read that 'rapid industrialisation was occurring in Israel where British industry can readily supply the necessary capital goods... But in the Arab world recent developments appear to confirm that the prospects for profitable economic dealings ... are at the best static and could in the long term decline.' Arms sales could evidently be a better proposition.

The long-term aim of a 'cabal', such as Gerald James has described in Britain, is the retention of power by a certain 'Establishment', likewise in the United States with the Bush-Saud relationship, and probably in Germany and France and maybe elsewhere. The 'Establishment' is not necessarily tied to a particular political party, although in Britain its links have always been with the Conservative Party and it has had close links with Republicans in the United States. In recent years, both New Labour in Britain and the Democrats in the United States have been associated with such 'cabals', while in France and Germany right-wing governments and even centre-left governments would be expected to be similarly associated. One can recall that Klaus Kinkel came from the BND (German Security Service) to be German Foreign Minister under Helmut Kohl's chancellorship, and that there has for long been an alliance of French diplomacy with the French Foreign Legion. So it seems that, as so often, it is impossible to understand economic relations without considering politics. And it apparently follows that politics means scandals and allegations of corruption – Kohl (Germany), Craxi (Italy), Mitterrand and Dumas (France), as well as Bush (USA) and Thatcher (UK) all bear witness.

Such connections can be no more than suggestive of an underlying plot, but it has been shown that Gerald James' story of 'cabals' does have strong credibility. He is able to quote a considerable body of literature in defence of his case including *The Silent Conspiracy* by Stephen Dorrill, and a most remarkable book, *Thatcher's Gold* by Paul

Halloran and Mark Hollingsworth, and there is Mark Phythian's *The Politics of British Arms Sales* (Manchester University Press, 2000). The conclusion must be that the economic relations between the West and the Islamic peoples of the 'Middle East' and North Africa, and now elsewhere in Africa, are not just simple economic market relations of mutual exchange of goods and services. There is an interdependent political and military relationship, which has existed now for a long time. There are some signs that this is weakening as it becomes subject to criticism both in the West and in the East. Arms flows to the region are being reduced. Oil supplies are increasingly coming from other oil fields. But few people accept Tony Blair's bland assertion that the second Iraq war was not about oil or that the nature of any new regime established in Iraq will not be influenced by the requirements of the giant world oil companies and their imperial masters.

There must be a rationale for the world-wide proliferation of United States military bases, with increasing numbers positioned in the 'Middle East' (Joseph Gerson & Bruce Burchard, *The Sun Never Sets: Confronting the Network of Foreign US Military Bases*). In addition to 700 US military bases already girdling the globe, the war in Afghanistan left behind US bases not only in that country but also in Uzbekistan and Kirghizstan in Central Asia. After the Iraq war it is said that there are to be 14 enduring military bases in the region. It is simply too difficult, if not impossible, to separate the political interest of the Western powers in the nature of the government regimes in the 'Middle East' from their economic interest in retaining control over the world's major reserves of oil and gas. As reserves of oil in Africa prove to be more and more important, great power interest in Africa has become evident. Both Mr Blair and President Bush have made recent visits to Africa. It would be nice to suppose that this is the result of concern for the AIDS epidemic sweeping across Africa, or the burden of debt to international financial institutions which afflicts African governments.

US interest in Africa was revealed in articles published by *Le Monde Diplomatique* (08.07.04). US bases are now established in São Tomé, conveniently in the centre of the West African oil states, and more recently in Djibouti across from the mouth of the Persian Gulf, through which flows much of the world's oil supplies, and just down the Red Sea from Port Sudan, the outlet for newly discovered Sudanese oil. Interest in military intervention on the Sudan has not only a humanitarian motive. Control over Sudan's oil has become an important issue between the Western powers. Oil production in the Sudan is currently being developed mainly by Chinese and French companies, and the French already have a military base in Djibouti. Apart from Nigeria, the oil producing states of West Africa are mainly one-time French colonies, where French was established as the common tongue. Inland from these coastal one-time colonies lies the Sahel. It is in this region south of the Sahara, with largely Islamic populations, that, according to *Le Monde Diplomatique*, US military activity has been greatest. General Charles Wald, Deputy US commander of EUROCOM, under which command Africa falls, commented after a visit in 2003 to Djibouti that 'the US and the French have common interests in the region'. He must have meant oil.

An organisation with the acronym of ACRI (African Crisis Response Initiative) was established by the Pentagon in 1997 as part of its Joint Army Training scheme for officers in friendly states to teach them NATO methods and the use of US military equipment. ACRI was changed by the Bush administration into ACOTA (African Contingency Operations Training Assistance) and to its remit of humanitarian and emergency operations 'offensive' training was added. The rationale for all this activity is said to be defence against terrorism, which, it is claimed, has only to be put down in one region, that it will spring up in another. A permanent centre for such training has been established at Abuja in Nigeria, and meetings of the military commands of Algeria, Chad, Mali, Mauritania, Morocco, Niger, Nigeria and Senegal have been held there under US auspices.

For all President George W Bush's talk of unilateralism, the search for military allies for the United States has evidently been a major pre-occupation of the administration – what Bush has called in relation to the war in Iraq 'the coalition of the willing'. These are those among the United Nations from the several oil states who have agreed, under various forms of duress, to finance the costs as in Iraq of US military adventures, to supply mercenaries to eke out the demands on US manpower, even to harbour US troops – Saudi Arabia yesterday,

Afghanistan, Uzbekistan, Kuwait and Iraq today, Sudan tomorrow. Nor does it seem from John Kerry's speeches that a Democratic administration would have behaved much differently, even while the unilateralism was less pronounced.

What is seen in the West, after September 11, 2001 in the United States and March 11, 2004 in Spain, as the threat of global terrorism to the West's whole social order has only served to reinforce the power and influence of the military in the US in particular. Chalmers Johnson in his book *The Sorrows of Empire* (Verso 2004) produces much evidence to show that what he calls 'the iron triangle' in the United States – of the Pentagon, the defence contractors, and the Armed Services Committee of Congress – has acquired power beyond even that of the President. American Presidents can no longer give orders but can only at best negotiate with the Pentagon. In his book *American Empire*, (Harvard University Press, 2003) Andrew J. Bacevich, an ex-serving US officer, explored this growing power over the years, citing the appointment as Secretary of State on his retirement from the army of Colin Powell, whom he describes as 'easily the most powerful JCS chairman in the history of that office,' and the elevation of the Commanders in Chief for each of the world's regions – Europe (NATO but also including Africa), Pacific (including East Asia), Central (Central Asia including the Persian Gulf) and South (Central and South America) – to coordinating power over all US armed forces in their region, and to supplanting the diplomatic corps in their rounds of visits. As Commander in Chief for Europe (CINCEUR), General Wesley Clark described their leading role as 'guiding Europe, the Middle East and Africa towards democracy, prosperity and stability for the coming decades.'

At the same time the proclaimed policy of the United States armed forces is that of full spectrum dominance – on land and sea, in the air, in space, and in the realm of information. This appears to be bi-partisan policy. There was no sign of the Democratic Party's aspirations for the Presidency abrogating this policy. Indeed, John Kerry presented himself and bowed out from his audiences with a military salute, and it was Mrs Albright, President Clinton's Secretary of State, who launched the war on Yugoslavia, demanding from General Powell, 'what was the use of having a great army which you are always boasting about, if we never employ it?' The central theme of Bacevich's book is that the United States has always been imperialist and he quotes as the epigraph to it the statement of President Theodore Roosevelt in December 1899, 'Of course, our whole national history has been one of expansion.' The long list of military actions by the United States since World War Two – in China, Korea, Guatemala, Indonesia, Cuba, the Congo, Peru, Laos, Vietnam, Cambodia, Grenada, Libya, El Salvador, Nicaragua, Panama, Iraq, Bosnia, Sudan, Yugoslavia, Afghanistan, Iraq again – occurred equally under Republican and Democratic administrations.

What has changed as a result of the rise to power under Bush Junior's presidency of the new Conservatives or Neocons, with their Project for the New American Century (PNAC) is the open adoption of a unilateralist posture. This unilateralism is an important change from the more consensual, albeit active, policy mode typified by President Carter's adviser Zbigniew Brzezinski, outlined in his book *The Choice*. This change has created serious problems for all America's allies throughout the world. Securing their compliance has raised big questions, which the horrific events in Palestine and Iraq every day make clear. The earlier assumption that the United States military dominance could achieve compliance is now open to question. Such overwhelming military power cannot get the lights and water supply working in Iraq, let alone the Iraq oil supplies flowing, which must have been a major aim of the US occupation. The idea of the new Bush administration that military power could win allies through the imposition of democratic regimes in other oil producing countries – Iran or Saudi Arabia – can hardly now be credible.

The crucial question now for Europe and equally for the Islamic peoples is what can be done to persuade the leading politicians of the United States, of whichever Party, to abandon their unilateralist military stance. In considering this question several key factors have to be taken into account. First, the huge United States economy is now largely dependent on imported supplies of fuel oil. Second, the sale of oil for dollar currency helps to make up for the increasing deficit on the US foreign balance of payments. Third, the world's reserves of easily extractable oil are running out, and the Islamic countries still hold the largest remaining reserves.

This change of US international policy has created serious problems not only for Europe, where most of the people have no desire to be dragged into wars behind the Anglo-American Alliance, but also for the United States itself. The cost of the Iraq war and its aftermath are putting a considerable strain on US resources in deficit at home and overseas to unprecedented amounts. An interesting indicator of the Bush cabal's methods for dealing with the Iraqi war debt was revealed in a story about the Iraqi government's inherited debt of some $200 billion, which appeared on *The Guardian* newspaper's front page on 13.10.04. This was that James Baker had been made special envoy of President Bush to collect Iraq's debts, including $27 billion owed to Kuwait. According to this story, which Naomi Klein had taken from documents obtained by *The Nation* magazine in New York:

> '$57 billion of Iraq debts would be assigned to a foundation created and controlled by a consortium in which the key players are the Carlyle group, where Baker is a senior counsellor, and the Albright Group (headed by another former US secretary of state, Madeleine Albright) and several other well connected firms. Under the deal, Kuwait would also give the consortium £2billion to invest in a private equity fund devised by the consortium with half of that going to Carlyle.'

Kathleen Clark, a law professor at Washington State University and a leading expert on government ethics and regulation, is reported to have commented that 'Mr Baker was in a classic conflict of interest... Even if Baker is somehow being screened from profiting from this deal, Carlyle is using Baker's government position to benefit themselves.' Carlyle, as was made clear earlier, is a giant US arms procurement company with close relations to both the Bush family and the Saud royal family. So, it is once again arms for oil.

In the United States such information as *The Nation* had obtained more easily comes to light than in the United Kingdom, where even two commissions of inquiry failed to obtain the full story of Mr Blair's connivance with President Bush over the Iraq war. Mr Blair as British Prime Minister with practically unchallenged control over Parliament had to reveal in defending his decision to go to war with Iraq that he was governed by the advice of the Joint Intelligence Committee, which was then shown to have been persuaded to 'sex up' the evidence. There can be no more important task for those concerned to work for peace and human rights than that of challenging the whole pervasive culture of government secrecy, legitimised in the name of national security against the threat of terrorism.

The problem posed equally for the West and the Islamic world is to find ways of arranging political-economic relations in such a way as to achieve a harmonious relationship. There are many questions to be answered. In particular would it be possible over time to diversify the currencies in which oil is traded, so as to reach a world-wide settlement embracing all the oil producing and consuming countries? What would need to be done to persuade American government and public opinion to accept a more sustainable long-term solution to the steady depletion of the planet's resources? What would be the benefits that could accrue from a new oil pricing policy to the governments and peoples of the oil producing countries – inevitably higher prices as the oil runs out, but in more diversified currencies? How can alternative energy resources be brought into the equation? How could reduction of arms sales be linked to an agreement on resource use? How could US backing for the apparently insatiable appetite of the Israelis for Palestinian land be overcome in such a settlement? Could a settlement of arms and oil relations make possible a major investment from the rich industrialised countries in the poor countries' economies and in the remission of their debts, such as could reduce the support for terrorist activities. The issue of arms for oil is such a large one, affecting the future of the planet itself, that the need for such a settlement has become of paramount importance.

APPENDIX TABLES

NOTE: All the tables that follow, even those from the UN organisations and from SIPRI and IISS, are taken from official published government sources. They should not be studied without heeding the warning that all governments' published statistics are frequently shown to be inaccurate, incomplete, and even deliberately distorted for political and/or security reasons. Only a few figures are available for years after 2001-2.

Table 1.
'Middle East' and North Africa, Population & GDP, 1975-2001

Country	Population (millions)		Urban %		GDP per head ($000s)	Highest	Growth Rates Pop. (%) GDP	
	1975	2001	1975	2001	2001	Year	2001/1975	
Qatar	0.3	0.7	79	92	28.1	2000	3.6	1.1
Israel	3.4	6.2	67	92	19.8	2000	2.3	2.0
Saudi Arabia	7.3	22.8	58	87	13.3	1980	4.4	-2.1
Oman	0.9	2.7	76	83	12.0	–	4.1	2.3
Libya	2.4	5.3	61	88	6.5	–	3.0	–
Tunisia	5.7	9.6	50	66	6.4	1986	2.0	2.0
Algeria	16.0	30.7	40	58	6.1	1985	2.5	-0.2
Iran	33.4	67.2	41	65	6.0	1976	2.7	-0.6
Turkey	41.0	69.3	42	66	5.9	1998	2.0	2.0
Lebanon	2.8	3.5	67	90	4.2	1998	0.9	4.0
Jordan	1.9	5.2	58	79	3.9	1986	3.8	0.3
Morocco	17.3	29.0	38	56	3.6	2001	2.0	1.3
Egypt	39.3	69.1	44	43	3.5	2001	2.2	2.8
Syria	7.5	17.0	45	52	3.3	1998	3.1	0.9
Sudan	16.7	32.2	19	37	2.0	2001	2.5	0.8
Palestine	1.3	3.3	60	–	1.3	–	3.7	–
Yemen	6.9	8.7	17	25	0.8	2001	3.8	–
TOTALS								
All Above	205	396						
Arab States	143	290	41	54	5.0	–	1.9	0.3

Source: UNDP Human Development Report 2003, Tables 5 & 12

Table 2.
Proved Fossil Reserves per Region in 1975
(000 m. tons of coal equivalent)

Region	Oil	Gas	Coal
North America	10.7	10.0	1500
Western Europe	6.0	6.0	120
Japan	0	0	20
Eastern Europe	1.4	0.6	100
USSR	17.0	24.0	4800
Oceania	0.5	1.8	27
China	4.5	0.7	1000
Far East	2.8	1.6	100
Middle East and North Africa	86.0	16.5	3
Sub-Saharan Africa	4.9	1.8	90
Latin America	7.2	4.0	40
World	141	67	7800
1975 Consumption	4	2	3

Source: P. Peeters, Can We Avoid a Third World War Around 2010?
Macmillan, 1979, Table 3.2

Table 3.
Major Oil Importers, 1987 and 1992

Country	Quantity (m.MT)		Value (US$b.)	
	1987	1992	1987	1992
USA	253	297	31	40
Japan	153	206	20	29
Germany (incl. East)	85	90	10	14
Italy	67	78	9	10
France	63	70	8	10
Korea	29	69	4	10
Netherlands	48	57	6	8
Spain	44	53	5	7
UK	32	48	4	7
Singapore	33	44	4	7
Brazil	31	24	4	3
Turkey	20	22	3	3
World	1,229	1,453	166	202

Source: *UNCTAD Commodity Yearbook 1994*

Table 4.
Major World Producers & Exporters of Oil, 1981 & 1992

Country	Oil Production (m.MT)		Oil Exports		
	1981	*1992*	*m.MT 1992*	*$b.1992*	*%*
World	2796	2947	1278	177	100
USA	422	360	–	–	
Canada	63	72	4	5	
UK	88	92	54	8	
Norway	23	103	93	13	
Australia	18	25	8	1	
USSR	609	447	54*	8	
China	101	139	22	3	
All Above	1264	1164	235	38	22
Argentina	26	27	–	–	
Mexico	120	134	71	9	
Venezuela	112	123	57	7	
Indonesia	79	67	38	5	
Malaysia	13	29	22	3	
India	15	28	–	–	
Nigeria	71	91	81	12	
Angola	7	26	20	3	
Gabon	8	15	13	2	
All Above	451	540	302	42	24
Bahrein	2	2	–	–	
Iran	72	171	115	16	
Iraq	44	2*	–	–	
Kuwait	57	53	25	2	
Oman	16	36	45	6	
Qatar	20	20	16	2	
S.Arabia	499	415	281	37	
Syria	8	23	9	1.5	
Turkey	2	4	–	–	
UAE	73	103	89	13	
Yemen	–	10	–	–	
Algeria	44	36	14	2	
Egypt	29	45	5	2	
Libya	58	69	–	–	
All west Asia & N. Africa	912	991	599	83	

Notes: All figures are of crude petroleum, not refined
      * = 144 tonnes in 1988 m.
      MT = million Metric Tonnes
Source: UNCTAD Commodity Yearbook 1994

45

Table 5.
'Middle East' & North Africa: Oil as % of All Exports, 2001

Country	All Exports ($ b.)	Oil Exports ($b.)	Proportion (%)
S. Arabia	76	70	90
Iran	26	23	89
Algeria	19	18	86
Kuwait	16	14	91
Qatar	12	10	89
Libya	10	9	94
Oman	10	8	80
Bahrein	8	5	67
Syria	5	4.5	76
UAE	9	4.3	51
Egypt	7	3	40
Yemen	4	4	95
All Above	202	173	86

Note: Oil Exports are of Crude Petroleum only
Source: UNCTAD, Handbook of Statistics 2003 Table 4.1A

Table 6.
Oil Prices, Gold Prices, Saudi Arabia's Surplus, 1979-2003

Year	Oil Price $ per barrel	Gold Price US$ per oz.	S.Arabia Surplus $ billion	Events
1979	18.7	512	14.6	Khomenei in Iran
1980	31.2	589	50.9	Iraq-Iran War
1981	35	397	49.3	
1982	31.4	380	9	Israel in Lebanon
1983	28.4	424	-20	
1984	28.3	360	-20	
1985	27	317	-15	
1986	13.8	567	-15	'Iran-gate' scandal
1987	17.8	446	-11	
1988	14.2	431	-8	
1989	17.2	381	-8	End of Iran war
1990	22	383	4	Iraq in Kuwait
1991	18.3	362	-14	End of Gulf war
1992	18.2	344	-9	
1993	16.1	360	-5	
1994	15.5	383	7	
1995	16.9	387	9	
1996	20.4	369	14	
1997	19.2	290	13	
1998	13.1	288	-9	East Asian crisis
1999	18.1	290	12	Yugoslav war
2000	28.2	274	30	US slump
2001	24.5	276	25	
2002	24.9	343	28	
2003	31	400	na	Iraq war

Sources: UNCTAD Commodity Yearbooks 1994 & 2003 and
IMF International Financial Statistics 1997 & 2003

Table 7.
West Asia and North Africa: Manufacturing Proportion of GDP, 1990, 1994 & 2001

Country	GDP ($b.) 1994	Mfg% GDP 1990	Mfg.% GDP 2001
Bahrein	4.1	17	12
UAE	5.4	10	14
Kuwait	24.3	12	7
Lebanon	–	11	8
Iran	63.7	12	16
Saudi Arabia	117	8	10
Syria	–	20	25
Turkey	74	20	15
Tunisia	81	17	18
Algeria	41.9	48	55
Jordan	6.1	15	15
Oman	11.6	4	8
Egypt	42.9	18	19
Morocco	30.8	18	17
Sudan	–	–	10
Yemen	–	9	7
Total/Average	506.8	16	15
East Asia Developing		20	25

Note: countries are listed in order of UNDP Human Development Index
Sources: UNCTAD *Statistical Yearbook 2003*, Table 7.3 and UNDP *Human Development Report 1997*, Table 26

Table 8.
'Middle East' and North Africa Ranking in Order of Manufacturing Value Added and Manufactured Exports, 1985 and 1998

Country	Manufacturing Value Added $ Per capita		Manufactured Exports $ Per capita	
	1985	*1998*	*1985*	*1998*
Switzerland	1	1	3	4
Singapore	16	4	1	1
Israel	23	22	18	17
Bahrein	25	27	23	36
Turkey	43	39	39	45
S.Arabia	33	42	30	35
Tunisia	54	49	–	–
Egypt	58	52	68	70
Oman	56	58	48	42
Morocco	62	59	49	56
Jordan	50	61	47	58
Algeria	42	64	34	49
Yemen	–	78	–	84
Nigeria	66	74	75	85

Source: UNIDO, *Industrial Development Report 2002/3*, Table A.2.14&15

Table 9.
'Middle East' & North Africa: Human Development, 2001

Country	HDIndex Max=1	Life Expect. at birth Years	Infant Mortality per1000	Adult Literacy %	Education Enrolment ratio	Under nourish %
Israel	0.9	78.9	6	95.1	90	–
Bahrein	0.84	73.7	13	87.9	81	–
Qatar	0.83	71.8	11	81.7	76	–
Kuwait	0.82	76.3	9	82.4	54	4
UAE	0.82	74.4	8	76.3	67	–
Libya	0.78	72.4	16	80.8	89	–
S.Arabia	0.77	71.9	23	77.1	58	3
Oman	0.75	72.2	12	73	58	–
Lebanon	0.75	73.3	28	86.5	76	3
Jordan	0.74	70.6	27	90.3	77	6
Tunisia	0.74	72.5	21	72.1	76	–
Turkey	0.73	70.1	36	85.5	60	–
Palestine	0.73	72.1	21	89.2	77	–
Iran	0.72	69.8	35	77.1	64	5
Algeria	0.7	69.2	39	67.8	71	6
Syria	0.68	71.5	23	75.3	59	3
Egypt	0.65	68.7	35	56.1	76	4
Morocco	0.61	68.1	39	49.8	51	7
Sudan	0.5	55.4	65	58.8	34	21
Yemen	0.47	59.4	79	47.7	52	33
Arab States Only						
2001	0.66	68	49	60.8	60	13
1970	0.61	66	128	50	–	16

Notes: HD Index is the UNDP combined index of human development Educational
enrolment ratio combines enrolment at all stages Under nourished refers to
children under 5

Sources: UNDP *Human Development Report 2000-1*, Tables 1&7
UNDP *Human Development Report 1997*, Tables 6&7

Table 10.
'Middle East' and North Africa: Expenditure in 1990 & 2001
On Education, Health & Military and Armed Forces numbers

Country	Education % of GDP 1990	2000	Health % of GDP 1990	2000	Military % of GDP 1990	2001	Armed Forces numbers, 000s 2001
Israel	6.3	7.3	3.8	8.3	12.2	7.7	1162
Bahrein	4.2	2.0	–	2.8	5.1	4.1	11
Qatar	3.5	3.6	–	2.5	–	–	12
Kuwait	4.8	–	4.0	2.6	48.5	11.3	16
UAE	1.9	1.9	0.8	2.5	4.7	2.5	42
S.Arabia	6.5	9.5	–	4.2	12.8	11.3	125
Oman	3.1	3.9	2.0	2.3	18.3	12.2	42
Lebanon	–	3.0	–	–	7.6	5.5	72
Turkey	2.2	3.5	2.2	3.6	3.5	4.9	515
Jordan	8.4	5.0	3.6	4.2	9.9	8.6	100
Iran	4.1	4.4	1.5	2.5	2.7	4.8	520
Algeria	5.3	–	3.0	3.0	1.5	3.5	137
Syria	4.1	4.1	0.4	1.6	6.9	6.2	319
Egypt	3.7	–	1.8	1.8	3.9	2.6	443
Morocco	5.3	5.5	0.9	1.3	4.1	4.1	196
Sudan	0.9	–	0.7	1.0	1.3	2.5	117
Yemen	–	10.0	1.1	–	8.5	6.1	67
All Average/Total	4.0	4.9	2.0	2.9	10.8	6.1	2896
USA	5.2	4.8	4.7	5.8	5.3	3.1	1414
UK	4.9	4.8	5.1	5.9	3.9	2.5	210

Notes: All figures are for public expenditure, excluding private
       The Gulf war distorts the 1990 military expenditure especially in Kuwait
Source: UNDP *Human Development Report 2003*, Table 17 & 20

Table 11.
'Middle East' & North Africa Defence Expenditure,
1985 & 2002

Country	Total US$b		$ per capita		as % of GDP	
	1985	*2002*	*1985*	*2002*	*1985*	*2002*
Bahrein	0.2	0.3	534	435	4.1	4.0
Egypt	4.6	3.1	95	44	13.0	3.9
Iran	7.9	4.9	178	67	7.7	4.6
Iraq	18.7	na	na	na	42.2	na
Israel	11.5	9.4	2709	1499	21.2	9.7
Jordan	0.9	0.8	261	162	15.9	9.3
Kuwait	2.2	3.3	1295	1582	9.1	10.7
Libya	1.2	0.5	311	98	6.2	3.8
Oman	2.4	2.6	1786	944	20.8	13.4
Qatar	0.3	1.7	824	2857	6.0	3.8
Saudi Arabia	31.5	21.0	2728	981	19.6	12.0
Syria	8.0	1.8	763	107	16.4	10.3
UAE	2.8	2.7	2003	949	7.6	4.0
Yemen	0.5	0.5	47	27	9.9	5.7
All Above	96	58	451	174	14.8	7.2
USA	381	330	1592	1138	6.1	3.3
NATO Europe	171	186	371	360	3.7	1.9
Soviet Union	368	(48)	1321	(333)	16.1	(4.8)
World	1171	843	243	136	6.2	2.6

Notes: ME&N.Africa accounted for 25% of world total in 1985, and for 18% in 2002
Figures in brackets are for Russia
Source: IISS, The Military Balance, 2003-2004

Table 12.
'Middle East' & North Africa: Arms 'Transfers'
Proportion of Total Imports, 1992 & 2002

Country	All Imports 1990 $m.	Arms Transfers 1992 $m. % of All 1990		Arms Transfers 2002 $m. % of All 1990	
Israel	16,794	1,330	0.8	226	0.1
Bahrein	3,712	35	0.9	51	1.4
Qatar	1,695	73	0.4	8	–
Kuwait	3,972	897	22.7	27	0.7
UAE	11,199	204	0.2	452	0.4
Saudi Arabia	24,069	1,198	5.0	478	2.0
Oman	1,681	20	0.1	48	0.3
Lebanon	2,525	132	0.5	4	–
Jordan	2,600	–	–	149	5.7
Tunisia	5,189	32	0.6	7	–
Turkey	22,302	1,347	6.1	721	4.5
Iran	15,716	386	2.4	298	1.8
Algeria	9,715	16	–	464	4.8
Syria	2,400	317	13.2	162	6.7
Egypt	7,862	995	12.8	638	8.1
Morocco	6,800	30	0.4	169	2.5
Yemen	1,476	–	–	496	33.6
Total	139,607	7,022	5.0	2,928	2.1

Note: ArmsTransfers are all adjusted to 1990 prices. Therefore, all percentages are related to the value of 1990 total imports.

Source: IMF *International Financial Statistics* and UNDP *Human Development Report* 2003, Table 20, based on SIPRI volume figures.

Table 13.
Arms Deliveries Proportion of World Trade, 1995-2002

Country/Trade	1995-8 Av. $b.	1999-2002 Av. $b.	1995-2002 Av. $b.	Arms %
Total Exports				
World	4900	3845	4370	100
USA	645	732	690	100
UK	214	273	240	100
Russia	82	98	90	100
Total Arms Exports				
World	45	35	40	0.9
USA	19	3.5	12	1.7
UK	14	5.5	10	4.2
Russia	3	4.5	4	4.4
'Middle East' Goods & Services Imports				
Israel	33	41	37	100
Saudi Arabia	45	50	47	100
Iran	16	16	16	100
Egypt	16	20	18	100
UAE	20	34	27	100
Kuwait	12	10	11	100
Total of Above	142	171	156	100
Arms Deliveries				
Israel	0.75	1.1	0.9	2.4
Saudi Arabia	9.5	6.7	8.0	17.0
Iran	0.52	1.7	1.1	2.3
Egypt	1.25	1.2	1.2	6.7
UAE	1.3	0.9	1.1	4.1
Kuwait	1.23	0.6	0.9	
Total of Above	14.6	11.0	12.5	8.0

Notes: 'Middle East' imports include goods and services
All Arms figures are at constant 2002 US$ prices
Grouped figures are all annual averages
Sources: IMF *International Financial Statistics 2003*
IISS, *The Military Balance 2003-4*, Table 37 & 39

Table 14.
Main World Arms Importers 1992-2002

Country	SIPRI Transfers $m			IISS Deliveries $m		
	1992	*2002*	*1998-2002*	*1995-8*	*1999-2002*	*2002*
Israel	1330	226	606	725	1080	700
Saudi Arabia	1198	478	890	9500	6650	5200
Egypt	995	638	650	1225	1170	2100
Kuwait	897	27	110	1225	600	1300
UAE	204	452	420	1275	900	900
Iran	386	298	360	525	175	na
Turkey	1347	721	936	na	na	na
Syria	317	162	130	100	100	na
Other ME & N.Africa	244	1243	756	1330	720	na
Total Above	6718	3043	4678	15,805	11,395	10,200
China	1163	2307	1760	na	na	1200
India	875	1668	963	na	na	900
S.Korea	497	229	689	na	na	600
Pakistan	261	1278	582	na	na	600
World Total	20,454	16,492	18,500	44,570	32,733	25,441

Notes: na = not available in this publication
Figures for several years are in all cases annual averages
SIPRI figures are $m at constant 1990 prices
IISS figures are $m at 2002 prices
Sources: for SIPRI, UNDP, *Human Development Report 2003*, Table 20
And SIPRI Yearbook, 2003, Table 13.1
For IISS, IISS *The Military Balance 2003-4* Tables on Pages 341 & 342

Table 15.
Main World Arms Exporters, 1998-2002

Country	SIPRI Transfers ($m.)		IISS Deliveries ($m.)	
	1998-2002	*2002*	*1995-2002*	*2002*
USA	7,500	3,940	16,140	10,240
UK	960	719	6,030	4,700
Russia	4,150	5,940	3,640	3,100
France	1,660	1,610	4,430	1,800
Germany	990	na	1,620	500
Italy	355	490	330	400
Other Europe	1,065	1,179	3,416	na
China	311	818	823	800
Others	1,475	1,206	2,359	260
TOTAL	18,508	16,406	823	800

Note: Grouped years = averages
SIPRI figures are all at 1990 prices
IISS figures are all at 2002 prices.
Sources: SIPRI figures from SIPRI Yearbook 2003 Table 13.1 and
UNCTAD, *Human Development Report 2003*, Table 20
IISS figures from IISS, *The Military Balance 2003-4* Tables on pp. 341-2

Table 16.
The World's Largest Petroleum and Arms Companies, 2000

Oil Companies	Country	Total Sales ($b.)
Exon Mobil	USA	206.1
Royal Dutch Shell	UK/Netherlands	184.6
BP	UK	149.1
Chevron-Texaco	USA	117.1
Total-Fina-Elf	France	105.8
Repsol YPF	Spain	42.6

Arms Companies		Arms Sales ($b.)
Lockheed Martin	USA	18.6
Boeing	USA	16.9
Raytheon	USA	14.0
BAE Systems	UK	13.2
General Dynamics	USA	6.5
Northrop Grumman	USA	6.7
EADS	France	4.6
Thales	France	4.3
United Technologies	USA	4.1
TRW	USA	4.0

Sources: UNCTAD *World Investment Report 2002*, Table IV.1
SIPRI Yearbook, Table 11.2

Table 17.
Saudi Arabia: Most Recent Arms Orders and Deliveries, 2001

Exporter	Classification	Designation	Number
Canada	Light armed Vehicle	LAV 25	1,117
UK	Ground Attack Fighter	Tornado IDS	48
France	Frigate with GM	Al Riyadh	3
USA	Military City/Port	Jizan	1
France	Helicopter	AS-532	12
USA	Airborne Warning/Control	AWACS E.3	5
Italy	Helicopter	AB-412TP	44
USA	Air to Air Missile	AMRAAM	475
USA	Anti-tank Guided Weapon	TOW 2A	1,827

Source: IISS The Military Balance, 2003-4, page 28

Table 18.
US Military Export Financing Programmes, 1992-2001
– all figures in $b.

Year	Grants	Loans	MAP	Debt Forgiven	Guarantee Reserve	Export/Import Bank
1992	3.93	0.34	0.12	0.08	0.03	0.03
1993	3.37	0.85	0.55	–	–	0.04
1994	3.15	0.77	0.32	0.09	0.08	
1995	3.15	0.36	0.12	–	–	0.02
1996	3.29	0.54	0.33	–	–	0.18
1997	3.23	0.3	0.07	–	–	0.02
1998	3.32	0.1	0.09	–	–	0.04
1999	3.37	–	0.27	–	–	0.32
2000	4.33	–	0.08	–	–	0.03
2001	3.53	–	0.04	–	0.03	0.02
Totals	34.58	3.47	2.00	0.18	0.03	0.98

Notes: MAP = Military Assistance Programme
Source: SIPRI Yearbook 2003, Table 13.1

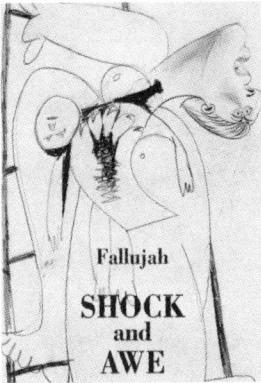